Name _____

Big Numbers

Periods →

Billions			Millions			Thousands			Ones		
100 billions	10 billions	billions	100 millions	10 millions	millions	100 thousands	10 thousands	thousands	hundreds	tens	ones
3	4	7	6	5	1	2	8	9	4	3	5

In a number, each group of 3 digits is separated by a comma. Each of these groups is called a period.

There are different ways to read and write numbers.

Standard form 347,651,289,435

Expanded form 300,000,000,000 + 40,000,000,000 + 7,000,000,000 + 600,000,000 + 50,000,000 + 1,000,000 + 200,000 + 80,000 + 9,000 + 400 + 30 + 5

Word form three hundred forty-seven billion, six hundred fifty-one million, two hundred eighty-nine thousand, four hundred thirty-five

Short word form 347 billion, 651 million, 289 thousand, 435

Write the numbers in **standard** form.

A. 60 thousand, 540 _____

B. seventy-two million six _____

C. 30,000 + 9000 + 100 + 7 _____

D. 46 billion, 32 thousand _____

E. 100,000,000 + 20,000,000 + 30,080 _____

Write the numbers in **expanded** form.

F. 76 thousand, 369 _____

G. 3 million, 425 _____

H. 69,005,002 _____

I. two billion, three thousand four _____

© Milliken Publishing Company   1   MP3495

Name _____   Comparing and ordering whole numbers

Number Workout

Write the number that comes **next**.

A.
4,657	12,320	84,308
4,658	12,321	84,309
_____	_____	_____

B.
107,200	283,998	492,206	99,998	999,998
107,201	283,999	492,207	99,999	999,999
_____	_____	_____	_____	_____

Write the number that comes **before**.

C.
_____	_____	_____	_____	_____
7,244	5,007	16,010	81,300	70,000
7,245	5,008	16,011	81,301	70,001

D.
_____	_____	_____	_____	_____
20,001	149,276	800,000	620,000	939,500
20,002	149,277	800,001	620,001	939,501

Write the numbers in order from the least to the greatest.

E. 43,091   43,901   43,019   43,910

_____ , _____ , _____ , _____

F. 712,674   712,476   712,746   712,467

_____ , _____ , _____ , _____

© Milliken Publishing Company   MP3495

Name _____   Rounding whole numbers

Rounding Numbers

Follow these steps to round numbers.

| 1. Find the place you want to round to. For example, suppose you want to round 32,654 to the nearest thousand. Look at the digit in the thousands place.

32,654
↑ | 2. Look at the digit to its right.

32,654
↑ | 3. If the digit is 5 or greater, round **up**. If it is less than 5, round **down**.

32,654
↑ greater than 5

32,654 rounded to the nearest thousand is 33,000 |

A. Round each number to the nearest **thousand.**

   8,431 _____   15,670 _____   38,910 _____

B. Round each number to the nearest **ten thousand.**

   45,244 _____   63,599 _____   3,848,125 _____

C. Round each number to the nearest **hundred thousand.**

   1,862,745 _____   6,457,208 _____

D. Round each number to the nearest **million.**

   5,381,262 _____   7,801,364 _____

Fun Fact! The earth's circumference (the distance around the earth at the equator) is 24,902 miles. If you traveled in a car at 60 miles per hour without stopping, it would take you a little over 17 days to drive that distance!

What is the earth's circumference rounded to the nearest hundred? _____

the nearest thousand? _____   the nearest ten thousand? _____

© Milliken Publishing Company   MP3495

Name _____   Exponents

Score With Exponents

Did you know that there is a shorter way to write 10 x 10 x 10 x 10? Just write **10⁴**. The 10 is called the **base**. It is the factor that is repeated. The 4 is called the **exponent**, or power. It tells how many times the base is used as a factor. Read 10⁴ as "ten to the fourth power."

Write the base and the exponent.

A. 4 x 4 x 4 _____   F. 8 x 8 _____

B. 2 x 2 x 2 x 2 _____   G. 1 x 1 x 1 x 1 x 1 _____

C. 6 x 6 x 6 _____   H. 5 x 5 x 5 x 5 _____

D. 10 x 10 x 10 x 10 x 10 _____   I. 7 x 7 x 7 _____

E. 3 x 3 x 3 x 3 _____   J. 9 x 9 x 9 x 9 x 9 _____

Write the numbers in standard form.

K. 2^3 __8__   P. 10^3 _____   U. 2^5 _____

L. 4^2 _____   Q. 1^8 _____   V. 3^3 _____

M. 3^4 _____   R. 2^2 _____   W. 10^6 _____

N. 5^2 _____   S. 4^3 _____   X. 5^3 _____

O. 2^6 _____   T. 2^7 _____   Y. 10^4 _____

© Milliken Publishing Company   MP3495

Exponents Express

This place value chart shows each place as a power of 10.

hundred thousands	ten thousands	thousands	hundreds	tens	ones
100,000	10,000	1,000	100	10	1
10 x 10 x 10 x 10 x 10	10 x 10 x 10 x 10	10 x 10 x 10	10 x 10	10	1
10^5	10^4	10^3	10^2	10^1	10^0

You can use powers of 10 when you write numbers in expanded form:

$4{,}253 = (4 \times 10^3) + (2 \times 10^2) + (5 \times 10^1) + (3 \times 10^0)$

Write each number in expanded form with exponents.

A. 3,296 _____

B. 4,715 _____

C. 63,502 _____

D. 84,790 _____

E. 99,000 _____

F. 720,463 _____

Write each number in standard form.

G. $(4 \times 10^5) + (1 \times 10^4)$ _____

H. $(5 \times 10^3) + (2 \times 10^2) + (7 \times 10^1) + (9 \times 10^0)$ _____

I. $(1 \times 10^4) + (7 \times 10^3) + (9 \times 10^2) + (6 \times 10^1) + (3 \times 10^0)$ _____

J. $(2 \times 10^4) + (8 \times 10^3) + (4 \times 10^2) + (1 \times 10^1) + (5 \times 10^0)$ _____

K. $(3 \times 10^5) + (9 \times 10^4) + (3 \times 10^2) + (6 \times 10^0)$ _____

Clipboard Math

Add or subtract.

A.	7,643 + 895	8,316 + 289	2,803 + 648	6,413 + 487
B.	6,075 − 546	5,319 − 744	2,605 − 135	3,259 − 608
C.	3,498 + 1,764	4,307 + 5,288	6,455 + 2,703	8,904 + 3,459
D.	4,600 − 3,208	2,912 − 1,761	7,954 − 4,287	9,168 − 4,289
E.	24,711 + 6,099	35,418 + 813	17,212 + 43,578	34,291 + 76,082
F.	53,243 − 647	60,000 − 1,264	35,488 − 12,882	71,351 − 65,847
G.	36,742 984 + 1,738	432 7,200 + 14,817	67,204 2,498 + 4,550	8,125 12,467 + 7,003

Name _____   Multiplying 3-digit numbers

Batty Over Multiplying

	1. Multiply by the ones digit.	2. Multiply by the tens digit.	3. Multiply by the hundreds digit.	4. Add the products.
156 x 234	156 x 234 **624**	156 x 234 624 **4,680**	156 x 234 624 4,680 **31,200**	156 x 234 **624** **4,680** **+ 31,200** **36,504**

Multiply.

A.	172 x 423	246 x 321	311 x 427	413 x 362	635 x 150
B.	203 x 277	489 x 333	376 x 21	757 x 35	943 x 18
C.	873 x 167	368 x 12	510 x 705	680 x 194	739 x 514

© Milliken Publishing Company

Name _____   One-digit divisors

Division Basics

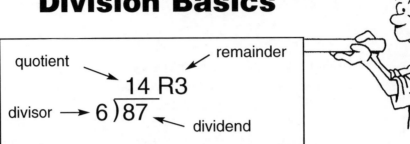

1. Write the first digit in the quotient.

$$\begin{array}{r}1\\6\overline{)87}\end{array}$$

There are enough tens to divide by 6. The first digit in the quotient goes above the tens (8).

2. Multiply.

$$\begin{array}{r}1\\6\overline{)87}\\6\end{array}$$

3. Subtract.

$$\begin{array}{r}1\\6\overline{)87}\\-6\\\hline 2\end{array}$$

4. Bring down the next digit in the dividend.

$$\begin{array}{r}1\\6\overline{)87}\\-6\downarrow\\\hline 27\end{array}$$

5. Repeat steps 1 to 4 until there are no numbers left to bring down.

$$\begin{array}{r}14\text{ R}3\\6\overline{)87}\\-6\\\hline 27\\-24\\\hline 3\end{array}$$

A. 3)97   5)82   6)95   7)147   8)249

B. 2)2,457   8)3,468   4)7,914   3)7,882   6)9,336

C. 5)59,316   9)12,854   6)81,767   3)64,128   8)24,896

Name _____ Zero in the quotient

Rockin' With Division

Sometimes when you divide, you need to write zero in the quotient.

$7\overline{)2{,}158}$

1. Divide hundreds.

$$\begin{array}{r} 3 \\ 7\overline{)2{,}158} \\ -\underline{2\,1} \\ 0 \end{array}$$

2. Bring down the tens. Divide tens.

$$\begin{array}{r} 30 \\ 7\overline{)2{,}158} \\ -\underline{2\,1}\downarrow \\ 05 \end{array}$$

There are not enough tens to divide by 7. Write 0 in the quotient over the tens (5).

3. Bring down the ones. Divide ones.

$$\begin{array}{r} 308\text{ R2} \\ 7\overline{)2{,}158} \\ -\underline{2\,1}\downarrow \\ 058 \\ -\underline{56} \\ 2 \end{array}$$

A. $3\overline{)618}$ $4\overline{)429}$ $2\overline{)620}$ $3\overline{)329}$ $4\overline{)835}$

B. $6\overline{)1{,}842}$ $8\overline{)3{,}277}$ $5\overline{)3{,}600}$ $4\overline{)2{,}006}$ $7\overline{)4{,}416}$

C. $2\overline{)16{,}083}$ $5\overline{)15{,}510}$ $9\overline{)18{,}128}$ $6\overline{)25{,}218}$ $3\overline{)21{,}286}$

© Milliken Publishing Company

Name _____   Two-digit divisors

Division Detective

Divide. Then check your answer by multiplying.

A.
```
        23 R1
    42)967         42
      -84         x 23
       127        126
      -126       +840
         1        966
                   +1
                  967
```
15)525

B. 28)1,960   62)1,615   70)2,739

C. 12)6,011   37)7,807   54)17,282

D. 18)7,660   99)11,899   82)19,188

Robot Mania

Multistep problems with whole numbers

Solve the problems by looking at each piece of information carefully. Answering the questions will help you.

A. A factory has been using robots for 3 years. This year it added 60 new robots. Now the factory has a total of 300 robots. Last year it doubled the number of robots from the first year. How many robots were used the first year?

1. How many robots are being used this year? _____
2. How many robots were used last year? _____
3. How many robots were used the first year? _____

B. Robot JR8 runs on 6 batteries. Each set of batteries lasts 16 hours. How many batteries are needed for Robot JR8 to work 192 hours?

1. How many hours does one set of 6 batteries last? _____
2. How many sets of batteries are needed for 192 hours? _____
3. How many batteries are needed for 192 hours? _____

C. Robot RG7A screws on 165 light bulbs in 20 minutes. How many light bulbs will be screwed on after 2 hours?

1. How many light bulbs are screwed on in 20 minutes? _____
2. How many 20-minute intervals make up 2 hours? _____
3. How many light bulbs are screwed on after 2 hours? _____

D. Robot ABC is made up of 135 more parts than Robot XYZ. Robot XYZ has 27 more parts than Robot QRS. Robot QRS is made up of 1,048 parts. How many parts does Robot ABC have?

1. How many parts does Robot QRS have? _____
2. How many parts does Robot XYZ have? _____
3. How many parts does Robot ABC have? _____

Name _____   Prime and composite numbers

Investigating Numbers

A **prime number** is a whole number greater than 1 whose only factors are 1 and itself.

The number 7 is a prime number.

1 x 7 = 7

The only factors of 7 are 1 and 7.

A **composite number** is a whole number greater than 1 that has more than two factors.

The number 8 is a composite number.

1 x 8 = 8   and   2 x 4 = 8

The factors of 8 are 1, 2, 4, and 8.

Look at each set of numbers. List the prime and composite numbers on the correct lines. Then shade in each box that has a prime number. You'll find a mystery letter!

A.

41	9	5
3	12	37
47	23	7

Prime _____

_____

Composite _____

_____

Mystery letter _____

B.

31	43	11
16	2	27
21	19	4

Prime _____

_____

Composite _____

_____

Mystery letter _____

C.

29	35	42
17	20	39
13	51	45

Prime _____

_____

Composite _____

_____

Mystery letter _____

Answer the questions.

D. What is the only even prime number? _____ Why are all other even numbers composite? _____

E. Why is the number 1 neither prime nor composite? _____

_____

© Milliken Publishing Company   MP3495

Name _____

Prime factorization

Prime Factors

A **factor tree** shows the prime factors of a number. Any composite number can be written as the product of prime factors. This is called **prime factorization**. Write the factors in order from the least to the greatest. Use exponents when you can. For example the prime factorization of 12 is $2^2 \times 3$.

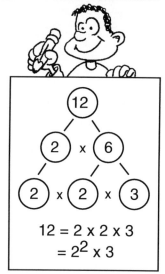

$12 = 2 \times 2 \times 3$
$= 2^2 \times 3$

Complete each factor tree. Then write the prime factorization.

A.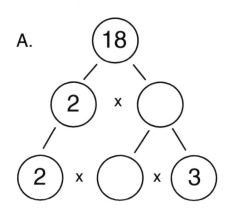

18 = _____

B.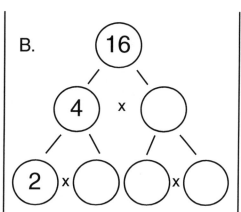

16 = _____

C.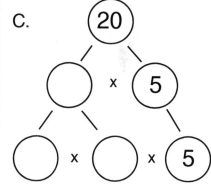

20 = _____

D.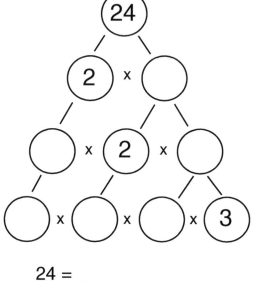

24 = _____

E.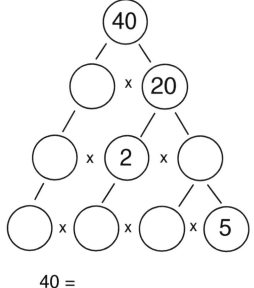

40 = _____

© Milliken Publishing Company

MP3495

Name _____   Greatest common factor

Factor Fun

Riddle: Why was the math book sad?

To find the answer to the riddle, first find the greatest common factor of each number pair below and write it in the box. Then use the code to help you write the matching letter for each factor you wrote. Write the letters on the lines.

Code
2 – I
3 – S
4 – M
5 – A
6 – L
7 – O
8 – R
9 – E
10 – Y
11 – N
12 – P
13 – H
15 – B
16 – D
20 – T

10, 14   40, 20   39, 13   15, 10   16, 48
□   □   □   □   □
___   ___   ___   ___   ___

21, 9   14, 7   12, 20   40, 15   22, 33   40, 30
□   □   □   □   □   □
___   ___   ___   ___   ___   ___

24, 36   16, 40   28, 21   30, 45   18, 24   27, 45   28, 16   12, 9
□   □   □   □   □   □   □   □
___   ___   ___   ___   ___   ___   ___   ___

14   MP3495

Name _____   Equivalent fractions

Picture Perfect

Two fractions that have the same value are called **equivalent fractions**. To find an equivalent fraction, multiply the numerator and denominator of a fraction by the same number.

$$\frac{3 \times 2}{4 \times 2} = \frac{6}{8}$$

$\frac{3}{4}$ and $\frac{6}{8}$ are equivalent fractions.

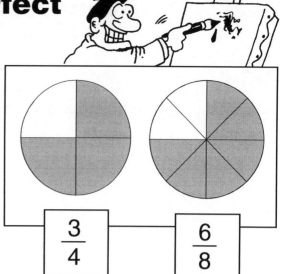

Write the missing numbers to make pairs of equivalent fractions.

A. $\frac{1}{2} = \frac{\square}{8}$ $\frac{3}{6} = \frac{9}{\square}$ $\frac{7}{12} = \frac{14}{\square}$

B. $\frac{2}{5} = \frac{10}{\square}$ $\frac{1}{8} = \frac{\square}{64}$ $\frac{4}{9} = \frac{12}{\square}$

C. $\frac{5}{6} = \frac{\square}{30}$ $\frac{3}{4} = \frac{\square}{12}$ $\frac{6}{7} = \frac{24}{\square}$

For each fraction, write two equivalent fractions.

D. $\frac{1}{4}$ _____ $\frac{1}{5}$ _____ $\frac{2}{9}$ _____

E. $\frac{2}{3}$ _____ $\frac{1}{6}$ _____ $\frac{3}{11}$ _____

Name _____ Simplifying fractions

Now Hear This!

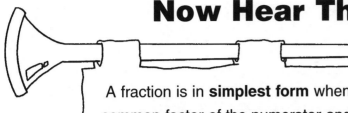

A fraction is in **simplest form** when the greatest common factor of the numerator and denominator is 1. To simplify a fraction, divide its numerator and denominator by a common factor. Keep dividing until the simplest form is reached.

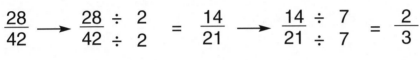

$\dfrac{28}{42}$ in simplest form is $\dfrac{2}{3}$

Write each fraction in simplest form.

A. $\dfrac{5}{10}$ _____ $\dfrac{9}{18}$ _____ $\dfrac{15}{20}$ _____

B. $\dfrac{2}{12}$ _____ $\dfrac{10}{25}$ _____ $\dfrac{16}{18}$ _____

C. $\dfrac{10}{40}$ _____ $\dfrac{9}{81}$ _____ $\dfrac{20}{30}$ _____

D. $\dfrac{24}{48}$ _____ $\dfrac{6}{10}$ _____ $\dfrac{15}{24}$ _____

E. $\dfrac{21}{28}$ _____ $\dfrac{12}{60}$ _____ $\dfrac{6}{18}$ _____

F. $\dfrac{10}{35}$ _____ $\dfrac{25}{30}$ _____ $\dfrac{12}{27}$ _____

© Milliken Publishing Company

Name _____

Improper fractions and mixed numbers

Fraction Pies

Some fractions are called **improper fractions**. In an improper fraction, the numerator is greater than the denominator.

$\dfrac{8}{4}$ and $\dfrac{7}{4}$ are improper fractions

Some improper fractions can be renamed as a whole number.

  $\dfrac{8}{4} = 2$

Some improper fractions can be renamed as a **mixed number**. A mixed number is written with a whole number and a fraction.

  $\dfrac{7}{4} = 1\dfrac{3}{4}$

Describe how much pie there is in each group below by writing both an improper fraction and a mixed number.

A.

B.

C.

Write each fraction as a whole number or a mixed number.

D. $\dfrac{9}{3}$ _____ $\dfrac{11}{4}$ _____ $\dfrac{7}{5}$ _____ $\dfrac{11}{2}$ _____

E. $\dfrac{17}{6}$ _____ $\dfrac{33}{8}$ _____ $\dfrac{29}{9}$ _____ $\dfrac{30}{5}$ _____

F. $\dfrac{20}{7}$ _____ $\dfrac{41}{10}$ _____ $\dfrac{18}{5}$ _____ $\dfrac{25}{8}$ _____

© Milliken Publishing Company

Name _____   Adding and subtracting fractions
                                                                 with like denominators

Can You Figure It Out?

Fractions that have the same denominators are said to have a **common denominator**. Add or subtract fractions with common denominators by adding or subtracting the numerators. Keep the denominator the same. Write your answer in simplest form.

$$\frac{3}{8} + \frac{1}{8} = \frac{4}{8} = \frac{1}{2}$$

$$\frac{11}{12} - \frac{7}{12} = \frac{5}{12}$$

$$\frac{3}{4} + \frac{2}{4} = \frac{5}{4} = 1\frac{1}{4}$$

Add or subtract. Write the answer in simplest form. If an answer is an improper fraction (greater than 1), write it as a whole number or mixed number.

A. $\quad \frac{2}{7} + \frac{1}{7} =$ $\qquad\qquad \frac{5}{9} + \frac{5}{9} =$ $\qquad\qquad \frac{3}{5} + \frac{6}{5} =$

B. $\quad \frac{10}{12} - \frac{3}{12} =$ $\qquad\qquad \frac{11}{4} - \frac{3}{4} =$ $\qquad\qquad \frac{9}{2} - \frac{8}{2} =$

C. $\quad \frac{2}{3} + \frac{7}{3} =$ $\qquad\qquad \frac{2}{8} + \frac{4}{8} =$ $\qquad\qquad \frac{7}{16} + \frac{11}{16} =$

D. $\quad \frac{12}{15} - \frac{9}{15} =$ $\qquad\qquad \frac{14}{6} - \frac{12}{6} =$ $\qquad\qquad \frac{9}{10} - \frac{5}{10} =$

E. $\quad \frac{8}{20} + \frac{5}{20} =$ $\qquad\qquad \frac{16}{24} + \frac{12}{24} =$ $\qquad\qquad \frac{11}{18} + \frac{9}{18} =$

Going Ape Over Fractions

To add or subtract fractions that have different denominators, first change the fractions to equivalent fractions that have the same denominator. Then add or subtract the numerators.

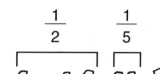

$$\frac{1}{2} + \frac{1}{5} = \frac{5}{10} + \frac{2}{10} = \frac{7}{10}$$

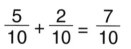

$$\frac{5}{10} + \frac{2}{10} = \frac{7}{10}$$

Add or subtract. Write the answer in simplest form.

A. $\dfrac{3}{8} + \dfrac{1}{4} =$

B. $\dfrac{2}{3} - \dfrac{1}{4} =$

C. $\dfrac{3}{5} + \dfrac{2}{10} =$

D. $\dfrac{11}{12} - \dfrac{3}{4} =$

E. $\dfrac{1}{3} + \dfrac{3}{7} =$

F. $\dfrac{9}{15} - \dfrac{3}{5} =$

G. $\dfrac{7}{12} + \dfrac{2}{6} =$

H. $\dfrac{5}{9} - \dfrac{1}{3} =$

I. $\dfrac{1}{4} + \dfrac{1}{5} =$

J. $\dfrac{7}{8} - \dfrac{3}{16} =$

K. $\dfrac{4}{7} + \dfrac{3}{14} =$

L. $\dfrac{9}{12} - \dfrac{1}{6} =$

Name _____   Adding and subtracting mixed numbers

High Flyers

When adding or subtracting mixed numbers, don't forget these rules.

- Use common denominators.
- Regroup if you need to.
- Write answers in simplest form.

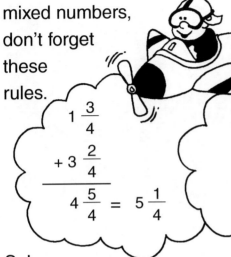

$1\frac{3}{4}$
$+3\frac{2}{4}$
$4\frac{5}{4} = 5\frac{1}{4}$

$7\frac{2}{3}$ → $7\frac{4}{6}$
$-4\frac{1}{6}$ → $-4\frac{1}{6}$
$3\frac{3}{6} = 3\frac{1}{2}$

$6\frac{1}{5}$ → $5\frac{6}{5}$
$-1\frac{3}{5}$ → $-1\frac{3}{5}$
$4\frac{3}{5}$

Solve.

A.
$6\frac{2}{5}$
$+2\frac{1}{5}$

$5\frac{3}{4}$
$+1\frac{1}{4}$

$2\frac{1}{10}$
$+3\frac{1}{5}$

$2\frac{4}{8}$
$+4\frac{1}{4}$

B.
$4\frac{2}{3}$
$-1\frac{1}{3}$

$7\frac{5}{6}$
$-2\frac{1}{2}$

$10\frac{3}{7}$
$-4\frac{6}{7}$

$8\frac{1}{5}$
$-7\frac{2}{10}$

C.
$6\frac{1}{2}$
$+6\frac{2}{5}$

$3\frac{9}{12}$
$+5\frac{4}{6}$

$1\frac{6}{15}$
$+8\frac{3}{5}$

$4\frac{5}{6}$
$+7\frac{3}{5}$

D.
$15\frac{3}{4}$
$-7\frac{1}{8}$

$12\frac{3}{4}$
$-9\frac{1}{2}$

$7\frac{5}{6}$
$-7\frac{3}{18}$

$8\frac{1}{5}$
$-3\frac{6}{10}$

© Milliken Publishing Company   MP3495

Let's Surf!

To multiply fractions, multiply the numerators and multiply the denominators.

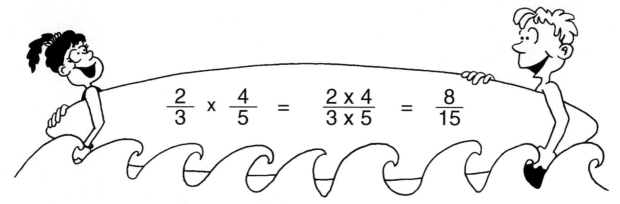

$$\frac{2}{3} \times \frac{4}{5} = \frac{2 \times 4}{3 \times 5} = \frac{8}{15}$$

Multiply. Write your answers in simplest form.

A. $\dfrac{1}{2} \times \dfrac{2}{5} =$ $\dfrac{3}{7} \times \dfrac{1}{4} =$ $\dfrac{5}{12} \times \dfrac{1}{3} =$

B. $\dfrac{1}{6} \times \dfrac{1}{7} =$ $\dfrac{4}{9} \times \dfrac{2}{3} =$ $\dfrac{2}{7} \times \dfrac{1}{2} =$

C. $\dfrac{5}{8} \times \dfrac{4}{5} =$ $\dfrac{2}{6} \times \dfrac{1}{3} =$ $\dfrac{1}{5} \times \dfrac{3}{4} =$

D. $\dfrac{1}{3} \times \dfrac{3}{5} =$ $\dfrac{5}{6} \times \dfrac{2}{10} =$ $\dfrac{3}{8} \times \dfrac{3}{5} =$

E. $\dfrac{1}{12} \times \dfrac{5}{6} =$ $\dfrac{2}{9} \times \dfrac{1}{5} =$ $\dfrac{7}{10} \times \dfrac{5}{6} =$

F. $\dfrac{2}{5} \times \dfrac{5}{8} =$ $\dfrac{7}{12} \times \dfrac{2}{4} =$ $\dfrac{1}{4} \times \dfrac{9}{10} =$

Name _____   Multiplying fractions—canceling

A Quick Trick

You can use **canceling** to make multiplying fractions easier and faster. To do this, divide the numerator of one fraction and the denominator of the other fraction by their greatest common factor.

$$\frac{5}{6} \times \frac{3}{7} \longrightarrow \frac{5}{\cancel{6}_2} \times \frac{\cancel{3}^1}{7} = \frac{5}{14}$$

The greatest common factor of 6 and 3 is 3. Divide both 6 and 3 by 3.

Multiply. Use canceling when you can. Write your answers in simplest form.

A. $\dfrac{3}{4} \times \dfrac{7}{12} =$

B. $\dfrac{2}{3} \times \dfrac{3}{5} =$

C. $\dfrac{5}{8} \times \dfrac{1}{3} =$

D. $\dfrac{1}{2} \times \dfrac{3}{4} =$

E. $\dfrac{12}{21} \times \dfrac{7}{9} =$

F. $\dfrac{5}{8} \times \dfrac{7}{10} =$

G. $\dfrac{5}{20} \times \dfrac{4}{5} =$

H. $\dfrac{2}{6} \times \dfrac{6}{7} =$

I. $\dfrac{4}{9} \times \dfrac{3}{5} =$

J. $\dfrac{3}{7} \times \dfrac{7}{9} =$

K. $\dfrac{7}{8} \times \dfrac{6}{7} =$

L. $\dfrac{5}{12} \times \dfrac{3}{20} =$

M. $\dfrac{7}{24} \times \dfrac{8}{14} =$

N. $\dfrac{3}{10} \times \dfrac{2}{18} =$

Cool Multiplying

When multiplying fractions with whole numbers or mixed numbers, change the whole number or mixed number to a fraction first.
Then multiply the same way you multiply fractions.

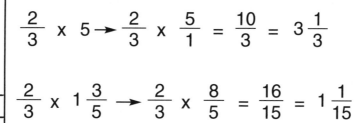

Multiply. Write your answers in simplest form.

A. $\dfrac{3}{4} \times 1\dfrac{1}{5} =$

B. $\dfrac{3}{5} \times 2\dfrac{1}{3} =$

C. $\dfrac{1}{5} \times 25 =$

D. $2\dfrac{5}{9} \times \dfrac{1}{7} =$

E. $\dfrac{1}{5} \times 3\dfrac{1}{4} =$

F. $1\dfrac{4}{6} \times \dfrac{1}{3} =$

G. $4 \times \dfrac{3}{8} =$

H. $\dfrac{11}{12} \times 2 =$

I. $\dfrac{1}{4} \times 2\dfrac{1}{3} =$

J. $\dfrac{3}{7} \times 10 =$

K. $\dfrac{7}{8} \times 3 =$

L. $\dfrac{1}{3} \times 3\dfrac{1}{3} =$

M. $2\dfrac{6}{9} \times \dfrac{1}{2} =$

N. $\dfrac{5}{6} \times 3\dfrac{3}{4} =$

Dividing by fractions

Building Math Skills

To divide by a fraction, first invert the divisor (flip the numerator and denominator). Then multiply.

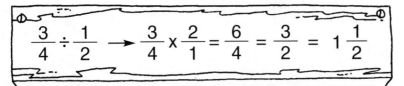

$\dfrac{3}{4} \div \dfrac{1}{2} \longrightarrow \dfrac{3}{4} \times \dfrac{2}{1} = \dfrac{6}{4} = \dfrac{3}{2} = 1\dfrac{1}{2}$

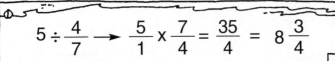

$5 \div \dfrac{4}{7} \longrightarrow \dfrac{5}{1} \times \dfrac{7}{4} = \dfrac{35}{4} = 8\dfrac{3}{4}$

$2\dfrac{3}{5} \div \dfrac{2}{5} \longrightarrow \dfrac{13}{\cancel{5}} \times \dfrac{\cancel{5}}{2} = \dfrac{13}{2} = 6\dfrac{1}{2}$

Divide.

A. $\dfrac{1}{6} \div \dfrac{3}{4} =$

B. $\dfrac{5}{7} \div \dfrac{1}{2} =$

C. $1\dfrac{3}{5} \div \dfrac{1}{5} =$

D. $5 \div \dfrac{1}{4} =$

E. $\dfrac{6}{9} \div \dfrac{5}{6} =$

F. $\dfrac{5}{8} \div \dfrac{1}{4} =$

G. $9 \div \dfrac{1}{3} =$

H. $3 \div \dfrac{2}{3} =$

I. $\dfrac{4}{9} \div \dfrac{1}{3} =$

J. $2\dfrac{1}{2} \div \dfrac{3}{8} =$

K. $\dfrac{3}{7} \div \dfrac{2}{7} =$

L. $8 \div \dfrac{2}{3} =$

M. $3\dfrac{1}{3} \div \dfrac{3}{6} =$

N. $\dfrac{11}{20} \div \dfrac{1}{5} =$

Stepping High

thousands	hundreds	tens	ones
1,000	100	10	1
			0

tenths	hundredths	thousandths
$\frac{1}{10}$	$\frac{1}{100}$	$\frac{1}{1,000}$
5	7	3

. (decimal point between ones and tenths)

Decimals express multiples of $\frac{1}{10}$ ($\frac{1}{10}$, $\frac{1}{100}$, $\frac{1}{1,000}$, and so on).

The value of a number to the right of a decimal point is less than 1.

$$0.573 = \frac{573}{1,000}$$

Riddle Fun!

How can you jump off a 50-foot ladder without getting hurt?

To find the answer, write the decimal for each numbered clue below. Then look at the letters next to the decimals you wrote. Write the letters on the matching lines at the bottom of the page.

1. 7 hundredths _____ F
2. 51 hundredths _____ G
3. $\frac{176}{1,000}$ _____ B
4. 25 thousandths _____ E
5. $\frac{9}{10}$ _____ T
6. 6 thousandths _____ J
7. 32 hundredths _____ R

8. 3 tenths _____ U
9. 8 tenths _____ O
10. $\frac{2}{1,000}$ _____ P
11. $\frac{1}{100}$ _____ N
12. $\frac{45}{100}$ _____ M
13. $\frac{604}{1,000}$ _____ H

__J__ __U__ __M__ __P__   __O__ __F__ __F__   __T__ __H__ __E__
0.006 0.3 0.45 0.002   0.8 0.07 0.07   0.9 0.604 0.025

__B__ __O__ __T__ __T__ __O__ __M__   __R__ __U__ __N__ __G__
0.176 0.8 0.9 0.9 0.8 0.45   0.32 0.3 0.01 0.51

Name _____   Rounding decimals

Decimal Roundup

Rounding decimals is a lot like rounding whole numbers. Look at the digit to the right of where you are rounding to. If the digit is 5 or greater, round up. If it is less than 5, round down.

Round to the nearest tenth.

A.	0.91 __**0.9**__		3.76 _____		15.42 _____
B.	2.37 _____		0.152 _____		0.803 _____
C.	1.06 _____		14.35 _____		7.52 _____
D.	0.136 _____		0.294 _____		0.673 _____

Round to the nearest hundredth.

E.	0.365 __**0.37**__		8.252 _____		0.619 _____
F.	5.073 _____		9.128 _____		4.355 _____
G.	0.876 _____		3.425 _____		12.091 _____
H.	7.387 _____		15.664 _____		41.503 _____

Round to the nearest whole number.

I.	3.5 __**4**__		6.81 _____		2.73 _____
J.	4.8 _____		9.53 _____		8.09 _____
K.	12.15 _____		75.55 _____		47.03 _____
L.	17.99 _____		29.84 _____		68.36 _____

© Milliken Publishing Company

A Decimal Lesson

Adding and subtracting decimals

When adding or subtracting decimals, first line up the decimal points. If you need to, add 0's so that there are the same number of digits to the right of decimal points. Then add or subtract.

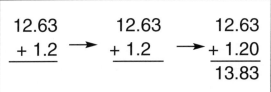

Adding 0's to the right of a decimal does not change its value.

Don't forget the decimal point in your answer.

```
12.63      12.63      12.63
+ 1.2  →   + 1.2  →  + 1.20
                     ─────
                      13.83
```

```
                             8 18
 9.862      9.862          9.8̸62
- 2.9      - 2.900        - 2.900
                          ──────
                           6.962
```

A.
5.3	1.83	7.6	12.75	3.5
+ 0.6	+ 2.5	+ 3.98	+ 5.4	+ 20.56

B.
12.9	6.8	41.9	16.74	25.1
− 7.4	− 5.6	− 2.85	− 0.7	− 8.39

C.
72.842	9.31	18.079	4.667	51.7
+ 6.003	+ 14.572	+ 16.2	+ 9.15	+ 8.999

D.
8.521	17.26	5.316	13.4	58.325
− 1.421	− 8.034	− 2.8	− 6.592	− 0.26

E.
20.11	6.01	8.2	4.25	3.052
6.7	9.532	48.7	0.609	7.5
+ 18.06	+ 14.9	+ 6.183	+ 7.4	+ 24.68

Name _____

Comparing decimals and fractions

Let's Compare

When comparing fractions and decimals, first express them in the same form. Then compare.

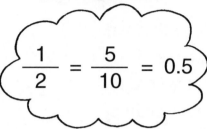
$\frac{1}{2} = \frac{5}{10} = 0.5$

Write each fraction as a decimal.

A. $\frac{7}{10}$ _____ $\frac{3}{10}$ _____ $\frac{8}{10}$ _____ $\frac{2}{10}$ _____

B. $\frac{49}{100}$ _____ $\frac{50}{100}$ _____ $\frac{9}{100}$ _____ $\frac{1}{100}$ _____

C. $\frac{623}{1,000}$ _____ $\frac{705}{1,000}$ _____ $\frac{3}{1,000}$ _____ $\frac{14}{1,000}$ _____

Write each decimal as a fraction. Write your answers in simplest form.

D. 0.3 _____   0.4 _____   0.5 _____

E. 0.79 _____   0.50 _____   0.25 _____

F. 0.281 _____   0.250 _____   0.015 _____

Compare the numbers. Write >, <, or = in each ◯.

G. $\frac{11}{20}$ ◯ 0.05 $\frac{1}{5}$ ◯ 0.05   0.8 ◯ $\frac{4}{5}$

H. 0.75 ◯ $\frac{15}{100}$ $\frac{9}{100}$ ◯ 0.9   0.15 ◯ $\frac{2}{10}$

I. $\frac{500}{1,000}$ ◯ 0.5   0.643 ◯ $\frac{65}{100}$ $\frac{15}{20}$ ◯ 0.721

© Milliken Publishing Company   MP3495

Multiplying Pointers

When multiplying decimals, the number of decimal places in the product must equal the total number of decimal places in the factors.

```
  0.6   ← 1 decimal place
x  4
  2.4   ← 1 decimal place
```

```
   0.3   ← 1 decimal place
x  0.2   ← 1 decimal place
  0.06   ← 2 decimal places
```
↖ Add 0's to the left of the product if there are not enough decimal places.

```
   2.15   ← 2 decimal places
x   1.3   ← 1 decimal place
   645
  2150
  2.795   ← 3 decimal places
```

Multiply.

A.
| 0.7 × 6 | 0.9 × 0.9 | 1.2 × 7 | 1.2 × 0.7 | 33 × 0.3 |

B.
| 3.26 × 0.4 | 0.15 × 0.3 | 8.12 × 0.4 | 152 × 0.4 | 3.07 × 0.5 |

C.
| 16 × 0.14 | 105 × 0.21 | 1.2 × 1.2 | 34 × 2.3 | 252 × 0.41 |

D.
| 2.4 × 3.1 | 0.24 × 3.1 | 15 × 0.017 | 1.3 × 0.22 | 4.6 × 3.4 |

© Milliken Publishing Company

MP3495

Name _____

Dividing decimals by whole numbers

Buzzin' With Division

To divide a decimal by a whole number, follow these steps.

| 1. Place a decimal point in the quotient directly above the decimal point in the dividend. $5\overline{)0.47}$ | 2. Divide as you would with whole numbers. If you need to, use 0 as a place holder. $\begin{array}{r} 0.09 \\ 5\overline{)0.47} \\ -45 \\ \hline 2 \end{array}$ | 3. Add 0's to the dividend if needed. Keep dividing until you can't divide anymore. $\begin{array}{r} 0.094 \\ 5\overline{)0.470} \\ -45\downarrow \\ \hline 20 \\ -20 \\ \hline 0 \end{array}$ | $5\overline{)0.47}$ |

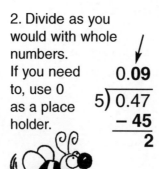
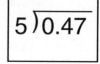

Divide.

A. $2\overline{)4.6}$ $5\overline{)52.5}$ $4\overline{)7.2}$ $3\overline{)20.4}$ $2\overline{)7.4}$

B. $6\overline{)4.8}$ $7\overline{)40.6}$ $2\overline{)2.46}$ $5\overline{)1.75}$ $4\overline{)8.64}$

C. $3\overline{)0.486}$ $2\overline{)3.92}$ $6\overline{)1.446}$ $3\overline{)8.643}$ $7\overline{)0.973}$

D. $15\overline{)9.45}$ $24\overline{)31.2}$ $36\overline{)86.4}$ $10\overline{)75.5}$ $41\overline{)2.091}$

© Milliken Publishing Company

MP3495

Hop to It!

Dividing decimals by decimals

To divide a decimal by a decimal, follow these steps.

1. Change the divisor to a whole number. Do this by multiplying it by a multiple of 10.

$$2.5\overline{)22.75} \rightarrow 25\overline{)22.75}$$
↑
Multiply by 10.

2. Multiply the dividend by the same number.

$$25\overline{)227.5}$$

3. Divide.

$$\begin{array}{r} 9.1 \\ 25\overline{)227.5} \\ -225 \\ \hline 25 \\ -25 \\ \hline 0 \end{array}$$

Multiply the divisor by 10, 100, or 1,000. Rewrite each problem. Then divide.

A. $0.3\overline{)1.2}$ $0.5\overline{)0.8}$ $0.03\overline{)0.153}$ $0.04\overline{)0.26}$ $0.007\overline{)0.028}$

$$\begin{array}{r} 4 \\ 3\overline{)12} \\ -12 \\ \hline 0 \end{array}$$

B. $2.1\overline{)4.83}$ $3.6\overline{)14.04}$ $1.5\overline{)0.045}$ $4.9\overline{)0.294}$ $0.05\overline{)0.21}$

C. $7.2\overline{)16.56}$ $0.63\overline{)4.41}$ $0.55\overline{)4.455}$ $0.009\overline{)0.054}$ $2.8\overline{)0.896}$

Name _____   Meaning of percent

Percent Puzzler

In the grid at the right, 45 out of 100 squares are shaded. You can compare the number of shaded squares to the total number of squares by writing $\frac{45}{100}$, or 45%. The symbol **%** stands for **percent**. Percent means *per hundred*.

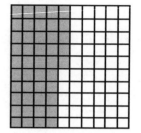

Write what percent of each grid is shaded.

A.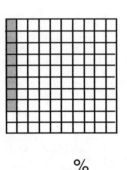
_____%

B.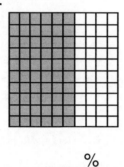
_____%

C.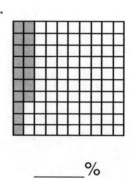
_____%

D.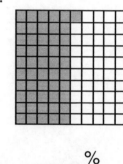
_____%

E.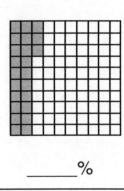
_____%

F.
_____%

G.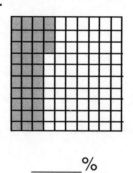
_____%

H.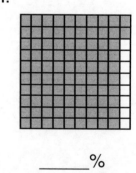
_____%

One hundred students were asked to name their favorite sports. Use the table to describe the results.

I. What percent of the students chose baseball? _____%
   What percent chose tennis? _____%

J. What percent of the students chose soccer or football? _____%

K. What percent of the students did not choose basketball? _____%

Favorite Sports of 100 Students	
Sport	Number of Students
Basketball	33
Baseball	18
Football	15
Soccer	21
Tennis	13

Name _____   Percents and fractions

What a Change!

You can change a percent into a fraction. Just use 100 as the denominator. Then if you need to, simplify the fraction.

You can change a fraction into a percent. First, write an equivalent fraction with a denominator of 100. Then change the fraction to percent.

$36\% = \dfrac{36}{100} = \dfrac{9}{25}$

$\dfrac{4}{5} = \dfrac{4 \times 20}{5 \times 20} = \dfrac{80}{100} = 80\%$

Write each percent as a fraction. Write your answers in simplest form.

A. 3% _____   7% _____   11% _____   50% _____

B. 10% _____   20% _____   15% _____   6% _____

C. 75% _____   25% _____   45% _____   70% _____

D. 54% _____   30% _____   80% _____   35% _____

Write each fraction as a percent.

E. $\dfrac{10}{100}$ _____ $\dfrac{1}{100}$ _____ $\dfrac{13}{100}$ _____ $\dfrac{49}{100}$ _____

F. $\dfrac{25}{50}$ _____ $\dfrac{10}{20}$ _____ $\dfrac{9}{10}$ _____ $\dfrac{99}{100}$ _____

G. $\dfrac{1}{50}$ _____ $\dfrac{4}{25}$ _____ $\dfrac{8}{20}$ _____ $\dfrac{2}{10}$ _____

H. $\dfrac{7}{20}$ _____ $\dfrac{3}{5}$ _____ $\dfrac{2}{50}$ _____ $\dfrac{21}{25}$ _____

© Milliken Publishing Company   MP3495

Name _____   Decimals and percents

Royal Math

You can write a percent as a decimal. First change the percent to a fraction with a denominator of 100. Then write the fraction as a decimal.

$$69\% = \frac{69}{100} = 0.69$$

You can write a decimal as a percent. First write the decimal as a fraction with a denominator of 100. Then write the fraction as a percent.

$$0.43 = \frac{43}{100} = 43\%$$

Write each percent as a decimal.

A. 9% _____
B. 15% _____
C. 48% _____
D. 1% _____
E. 37% _____
F. 52% _____
G. 66% _____
H. 25% _____
I. 74% _____
J. 89% _____
K. 4% _____
L. 93% _____

Write each decimal as a percent.

M. 0.38 _____
N. 0.99 _____
O. 0.72 _____
P. 0.44 _____
Q. 0.16 _____
R. 0.83 _____
S. 0.03 _____
T. 0.07 _____
U. 0.5 _____
V. 0.8 _____
W. 0.02 _____
X. 0.2 _____

Riddle Fun! Why is a book like a king?

To find out, look at the answers you wrote. Find their match below. Write the matching letters on the correct lines. (You will not use all of your answers.)

___ ___ ___ ___   ___ ___ ___ ___
0.15 72% 7% 0.25   0.25 0.09 80% 0.37

     ___ ___ ___ ___ ___
     44% 0.09 0.66 0.37 3%

© Milliken Publishing Company   34   MP3495

Name _____   Percent of a number

Flower Power

Mrs. Green has 80 potted plants in her flower shop. If 25% of her pots held violets, how many pots would that be?

Here are two ways you can find the percent of a number.

A. Write the percent as a decimal. Then multiply.

$25\% = 0.25 \rightarrow$
```
  0.25
× 80
-----
 20.00
```

Mrs. Green had 20 pots that held violets.

B. Write the percent as a fraction. Then multiply.

$25\% = \dfrac{25}{100} = \dfrac{1}{4} \rightarrow \dfrac{1}{4} \times 80 = 20$

Mrs. Green had 20 pots that held violets.

Solve. Write your answers in the flowers.

A. 50% of 30   B. 20% of 65   C. 80% of 20   D. 25% of 36

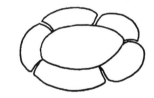 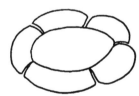  

E. 70% of 70   F. 75% of 40   G. 10% of 600   H. 15% of 800

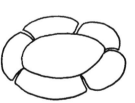 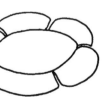 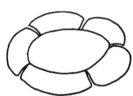 

I. 45% of 400   J. 17% of 300   K. 35% of 960   L. 52% of 750

Name _____   Angle measures

Super Sleuth

Help Sherlock figure out the missing angle measures.
Remember: The 3 angles of a triangle add up to 180°.
The 4 angles of a quadrilateral add up to 360°.

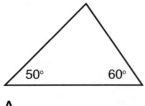   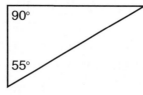   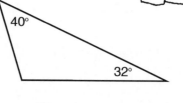

A. _____   B. _____   C. _____

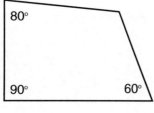   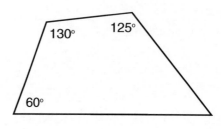

D. _____   E. _____

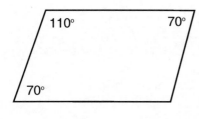   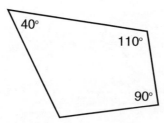   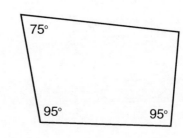

F. _____   G. _____   H. _____

Take the Super Sleuth Challenge!

1. Can a triangle have two angles that measure 90°? _____ Why or why not?
   _____

2. Can a quadrilateral have two angles that measure 90°? _____ Why or why not? _____

© Milliken Publishing Company

Name _____   Perimeter and area of complex figures

Floor Plan Puzzles

Mr. Wood is a builder. Look at the floor plan of the room he is planning to build. Figure out the perimeter and area of the room.

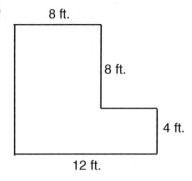

Perimeter — Figure out the lengths of all the sides. Then find their sum.

Perimeter of figure = 12 + 4 + 4 + 8 + 8 + 12
= 48 ft.

Area — Divide the room into simple figures. Find the area of each figure. Then find their sum.

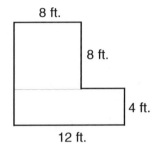

Area of square = 8^2 = 64 sq. ft.
Area of rectangle = 12 x 4 = 48 sq. ft.
Area of figure = 64 + 48 = 112 sq. ft.

Find the perimeter and area of each figure below.

A.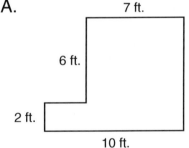

P = _____
A = _____

B.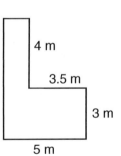

P = _____
A = _____

C.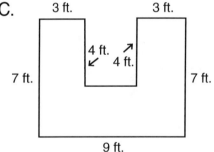

P = _____
A = _____

D.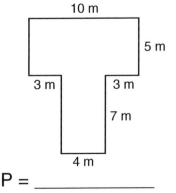

P = _____
A = _____

E.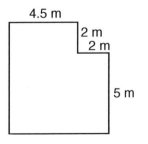

P = _____
A = _____

F.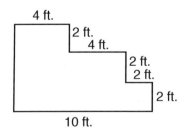

P = _____
A = _____

Name _____ Area—triangle and parallelogram

Land for Sale

Farmer Sprouts wants to buy some land. Look at the plots of land below. Write their areas. Farmer Sprouts will buy the plot of land that has the greatest area.

(Note: The figures are not drawn to scale.)

Area of triangle
$= \frac{1}{2} \times base \times height$

Area of parallelogram
$= base \times height$

A.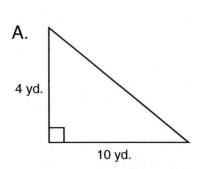

Area = _____ sq. yd.

B.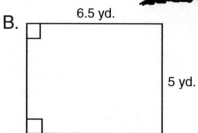

Area = _____ sq. yd.

C.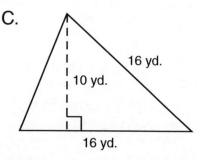

Area = _____ sq. yd.

D.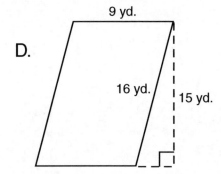

Area = _____ sq. yd.

E.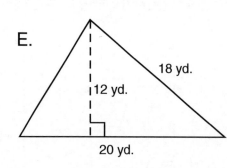

Area = _____ sq. yd.

F.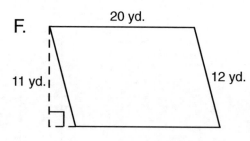

Area = _____ sq. yd.

G.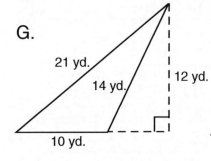

Area = _____ sq. yd.

H.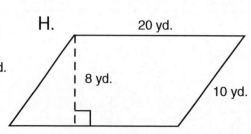

Area = _____ sq. yd.

I.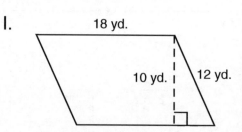

Area = _____ sq. yd.

Which plot of land will Farmer Sprouts buy? _____

© Milliken Publishing Company

Name _____   Surface area of rectangular prisms

Painting Challenge

Jinni is going to paint some blocks of wood. She wants to find out which block will need the most paint. Help her out by finding the **surface area** of each block.

A block, or rectangular prism, has a top and bottom that are the same size, a front and back that are the same size, and two side panels that are the same size. To find the surface area, find the area of the top (or bottom), the front (or back), and one side. Then multiply the sum of the areas by 2.

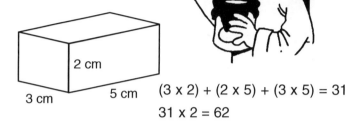

(3 x 2) + (2 x 5) + (3 x 5) = 31
31 x 2 = 62

The surface area of this block is 62 square cm.

Write the surface area of each figure. (Note: The figures are not drawn to scale.)

A.
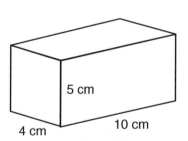
_____ sq. cm

B.
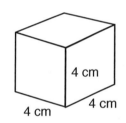
_____ sq. cm

C.
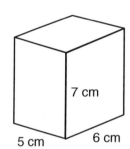
_____ sq. cm

D.
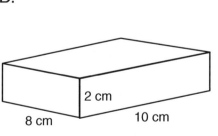
_____ sq. cm

E.
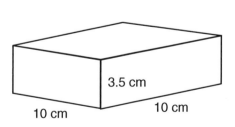
_____ sq. cm

F.
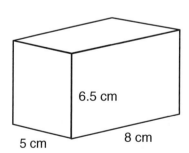
_____ sq. cm

Which block has the greatest surface area and will need the most paint? _____

© Milliken Publishing Company   MP3495

Name _____

Volume—rectangular prisms

Fill the Boxes

Lisa wants to measure the **volume** of her plastic boxes. She's going to fill each box with one-inch cubes. She's figured out the volume of one box already. Help her find the volume of the rest of her boxes.

(Note: The figures are not drawn to scale.)

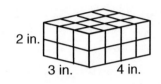

4 x 3 x 2 = 24

The volume of this box is 24 cubic inches.

Volume of rectangular prism
= *length* x *width* x *height*

A.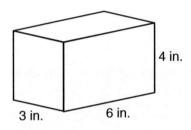

_____ cu. in.

B.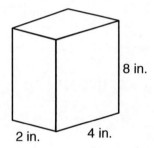

_____ cu. in.

C.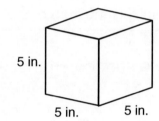

_____ cu. in.

D.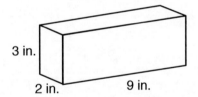

_____ cu. in.

E.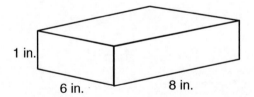

_____ cu. in.

F.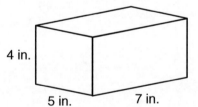

_____ cu. in.

G.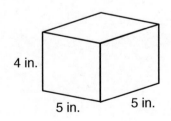

_____ cu. in.

H.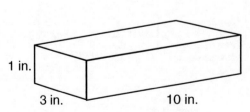

_____ cu. in.

I.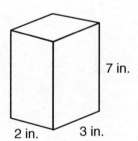

_____ cu. in.

© Milliken Publishing Company

MP3495

Name _____

Date _____ Score _____

Assessment A
Number Sense

Shade in the circle of the correct answer.

Find the matching number.

1. three million, six hundred five
   - ○ 300,605
   - ○ 3,000,605
   - ○ 3,000,650

2. 700,000 + 200 + 40 + 3
   - ○ 700,243
   - ○ 702,430
   - ○ 720,403

Find the number that comes before.

3. 2,163,000
   - ○ 2,163,999
   - ○ 2,163,001
   - ○ 2,162,999

4. 75,539,200
   - ○ 75,538,999
   - ○ 75,539,199
   - ○ 75,540,000

Find the matching number.

5. 618.523 rounded to the nearest tenth
   - ○ 618.53
   - ○ 618.5
   - ○ 618.522

6. 735.739 rounded to the nearest hundredth
   - ○ 735.7
   - ○ 735.8
   - ○ 735.74

Solve.

7. 2^5
   - ○ 32   ○ 19   ○ 10

8. 4×10^3
   - ○ 403   ○ 400   ○ 4,000

9. $(8 \times 10^4) + (9 \times 10^2)$
   - ○ 8,900   ○ 80,900   ○ 80,090

10. $(3 \times 10^5) + (6 \times 10^4) + (5 \times 10^0)$
    - ○ 365,000   ○ 36,050   ○ 360,005

© Milliken Publishing Company

Name _____
Date _____ Score _____

Assessment B
Computation

Shade in the circle of the correct answer.

Solve.

1. 65,314
 + 39,088
 ○ 94,402
 ○ 104,402
 ○ 104,492

2. 3,000,000
 − 167,009
 ○ 2,832,991
 ○ 2,932,991
 ○ 3,932,001

3. 321
 × 421
 ○ 130,001
 ○ 135,141
 ○ 135,101

4. A bolt of cloth is 3,185 feet long. If it is cut into pieces that are 13 feet long, how many pieces would there be?
 ○ 2.4   ○ 245   ○ 145

5. 4.26
 × 0.3
 ○ 1,278
 ○ 12.78
 ○ 1.278

6. 0.05)̄0.21
 ○ 4.2
 ○ 42
 ○ 4.02

7. $7\frac{1}{3}$
 $-4\frac{3}{4}$
 ○ $2\frac{2}{7}$
 ○ $3\frac{1}{4}$
 ○ $2\frac{7}{12}$

8. $\frac{5}{8} \times \frac{7}{10} =$
 ○ $\frac{7}{16}$ ○ $\frac{12}{18}$ ○ $\frac{12}{80}$

9. $\frac{11}{20} \div \frac{2}{5} =$
 ○ $1\frac{3}{8}$ ○ $\frac{22}{100}$ ○ $1\frac{5}{40}$

10. $3\frac{1}{3} \div \frac{3}{6} =$
 ○ $\frac{5}{3}$ ○ $6\frac{2}{3}$ ○ $3\frac{1}{3}$

© Milliken Publishing Company   MP3495

Name _____

Date _____ Score _____

**Assessment C
Fractions, Decimals,
and Percents**

  Shade in the circle of the correct answer.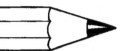

Find the equivalent decimal.

1. $\frac{16}{1,000}$ ○ 1.6   ○ 0.16   ○ 0.016

2. 4% ○ 0.04   ○ 4.0   ○ 0.4

3. $\frac{4}{5}$ ○ 0.5   ○ 0.4   ○ 0.8

Find the equivalent fraction.

4. 5% ○ $\frac{5}{10}$ ○ $\frac{1}{20}$ ○ $\frac{50}{100}$

5. 0.25 ○ $\frac{25}{10}$ ○ $\frac{1}{2}$ ○ $\frac{1}{4}$

6. 0.08 ○ $\frac{8}{10}$ ○ $\frac{2}{25}$ ○ $\frac{4}{100}$

Find the equivalent percent.

7. 0.39   3.9% 39%   0.39%
             ○   ○   ○

8. 0.9   90%   9%   0.9%
             ○   ○   ○

9. $\frac{17}{20}$ 85%   17%   34%
             ○   ○   ○

Find the missing sign.

10. 0.15 ◯ $\frac{2}{10}$ >   <   =
                              ○   ○   ○

11. $\frac{3}{5}$ ◯ 0.6   >   <   =
                              ○   ○   ○

12. 0.763 ◯ $\frac{75}{100}$ >   <   =
                              ○   ○   ○

Solve.

13. 25% of 40
    10   4   8
    ○   ○   ○

14. 70% of 140
    98   9.8   35
    ○   ○   ○

15. 52% of 500
    250   260   300
    ○   ○   ○

© Milliken Publishing Company

MP3495

Name _____

Date _____ Score _____

Assessment D
Geometry and
Measurement

Shade in the circle of the correct answer.

Find the missing angle measures.

1.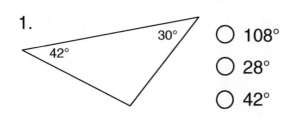
   ○ 108°
   ○ 28°
   ○ 42°

2.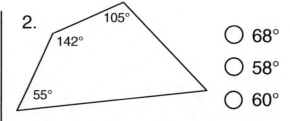
   ○ 68°
   ○ 58°
   ○ 60°

Find the area.

3.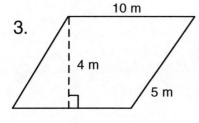
   ○ 40 sq. m
   ○ 20 sq. m
   ○ 50 sq. m

Find the area.

4.
   ○ 57.5 sq. ft.
   ○ 15 sq. ft.
   ○ 22.5 sq. ft.

Find the perimeter.

5.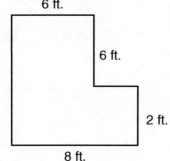
   ○ 22 ft.
   ○ 32 ft.
   ○ 30 ft.

Find the area.

6.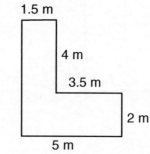
   ○ 14.5 sq. m
   ○ 16 sq. m
   ○ 20 sq. m

Find the surface area.

7.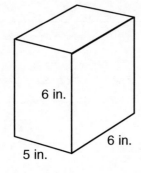
   ○ 192 sq. in.
   ○ 180 sq. in.
   ○ 360 sq. in.

Find the volume.

8.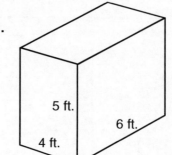
   ○ 20 cu. ft.
   ○ 120 cu. ft.
   ○ 60 cu. ft.

© Milliken Publishing Company

MP3495

Answer Key

Page 1
A. 60,540
B. 72,000,006
C. 39,107
D. 46,000,032,000
E. 120,030,080
F. 70,000 + 6,000 + 300 + 60 + 9
G. 3,000,000 + 400 + 20 + 5
H. 60,000,000 + 9,000,000 + 5,000 + 2
I. 2,000,000,000 + 3,000 + 4

Page 2
A. 4,659; 12,322; 84,310
B. 107,202; 284,000; 492,208; 100,000; 1,000,000
C. 7,243; 5,006; 16,009; 81,299; 69,999
D. 20,000; 149,275; 799,999; 619,999; 939,499
E. 43,019; 43,091; 43,901; 43,910
F. 712,467; 712,476; 712,674; 712,746

Page 3
A. 8,000; 16,000; 39,000
B. 50,000; 60,000; 3,850,000
C. 1,900,000; 6,500,000
D. 5,000,000; 8,000,000

Fun Fact! 24,900; 25,000; 20,000

Page 4
A. 4^3
B. 2^4
C. 6^3
D. 10^5
E. 3^4
F. 8^2
G. 1^5
H. 5^4
I. 7^3
J. 9^5
K. 8
L. 16
M. 81
N. 25
O. 64
P. 1,000
Q. 1
R. 4
S. 64
T. 128
U. 32
V. 27
W. 1,000,000
X. 125
Y. 10,000

Page 5
A. $(3 \times 10^3) + (2 \times 10^2) + (9 \times 10^1) + (6 \times 10^0)$
B. $(4 \times 10^3) + (7 \times 10^2) + (1 \times 10^1) + (5 \times 10^0)$
C. $(6 \times 10^4) + (3 \times 10^3) + (5 \times 10^2) + (2 \times 10^0)$
D. $(8 \times 10^4) + (4 \times 10^3) + (7 \times 10^2) + (9 \times 10^1)$
E. $(9 \times 10^4) + (9 \times 10^3)$
F. $(7 \times 10^5) + (2 \times 10^4) + (4 \times 10^2) + (6 \times 10^1) + (3 \times 10^0)$
G. 410,000
H. 5,279
I. 17,963
J. 28,415
K. 390,306

Page 6
A. 8,538; 8,605; 3,451; 6,900
B. 5,529; 4,575; 2,470; 2,651
C. 5,262; 9,595; 9,158; 12,363
D. 1,392; 1,151; 3,667; 4,879
E. 30,810; 36,231; 60,790; 110,373
F. 52,596; 58,736; 22,606; 5,504
G. 39,464; 22,449; 74,252; 27,595

Page 7
A. 72,756; 78,966; 132,797; 149,506; 95,250
B. 56,231; 162,837; 7,896; 26,495; 16,974
C. 145,791; 4,416; 359,550; 131,920; 379,846

Page 8
A. 32 R1; 16 R2; 15 R5; 21; 31 R1
B. 1,228 R1; 433 R4; 1,978 R2; 2,627 R1; 1,556
C. 11,863 R1; 1,428 R2; 13,627 R5; 21,376; 3,112

Page 9
A. 206; 107 R1; 310; 109 R2; 208 R3
B. 307; 409 R5; 720; 501 R2; 630 R6
C. 8,041 R1; 3,102; 2,014 R2; 4,203; 7,095 R1

Page 10
A. 23 R1; 35
B. 70; 26 R3; 39 R9
C. 500 R11; 211; 320 R2
D. 425 R10; 120 R19; 234

Page 11
A. 1. 300 robots
   2. 240 robots (300 − 60)
   3. 120 robots (240 ÷ 2)
B. 1. 16 hours
   2. 12 sets (192 ÷ 16)
   3. 72 batteries (12 x 6)
C. 1. 165 light bulbs
   2. 6 intervals
   3. 990 light bulbs (6 x 165)
D. 1. 1,048 parts
   2. 1,075 parts (1,048 + 27)
   3. 1,210 parts (1,075 + 135)

Page 12
A. Prime: 41, 5, 3, 37, 47, 23, 7
   Composite: 9, 12
   Mystery letter: U
B. Prime: 31, 43, 11, 2, 19
   Composite: 16, 27, 21, 4
   Mystery letter: T
C. Prime: 29, 17, 13, 51
   Composite: 35, 42, 20, 39, 45
   Mystery letter: L
D. 2; All other even numbers have at least 3 factors: 1, itself, and 2.
E. The number 1 has only 1 factor.

Page 13

A. $18 = 2 \times 3^2$
B. $16 = 2^4$
C. $20 = 2^2 \times 5$
D. $24 = 2^3 \times 3$
E. $40 = 2^3 \times 5$

Page 14

Answer to riddle: IT HAD SO MANY PROBLEMS

Page 15

A. $\frac{4}{8}$, $\frac{9}{18}$, $\frac{14}{24}$
B. $\frac{10}{25}$, $\frac{8}{64}$, $\frac{12}{27}$
C. $\frac{25}{30}$, $\frac{9}{12}$, $\frac{24}{28}$

Answers will vary. Examples:

D. $\frac{2}{8}$, $\frac{3}{12}$; $\frac{3}{15}$, $\frac{4}{20}$; $\frac{4}{18}$, $\frac{6}{27}$
E. $\frac{6}{9}$, $\frac{8}{12}$; $\frac{2}{12}$, $\frac{3}{18}$; $\frac{6}{22}$, $\frac{9}{33}$

Page 16

A. $\frac{1}{2}$, $\frac{1}{2}$, $\frac{3}{4}$
B. $\frac{1}{6}$, $\frac{2}{5}$, $\frac{8}{9}$
C. $\frac{1}{4}$, $\frac{1}{9}$, $\frac{2}{3}$
D. $\frac{1}{2}$, $\frac{3}{5}$, $\frac{5}{8}$
E. $\frac{3}{4}$, $\frac{1}{5}$, $\frac{1}{3}$
F. $\frac{2}{7}$, $\frac{5}{6}$, $\frac{4}{9}$

Page 17

A. $\frac{9}{4}$, $2\frac{1}{4}$
B. $\frac{11}{3}$, $3\frac{2}{3}$
C. $\frac{9}{2}$, $4\frac{1}{2}$
D. 3, $2\frac{3}{4}$, $1\frac{2}{5}$, $5\frac{1}{2}$
E. $2\frac{5}{6}$, $4\frac{1}{8}$, $3\frac{2}{9}$, 6
F. $2\frac{6}{7}$, $4\frac{1}{10}$, $3\frac{3}{5}$, $3\frac{1}{8}$

Page 18

A. $\frac{3}{7}$, $1\frac{1}{9}$, $1\frac{4}{5}$
B. $\frac{7}{12}$, 2, $\frac{1}{2}$
C. 3, $\frac{3}{4}$, $1\frac{1}{8}$
D. $\frac{1}{5}$, $\frac{1}{3}$, $\frac{2}{5}$
E. $\frac{13}{20}$, $1\frac{1}{6}$, $1\frac{1}{9}$

Page 19

A. $\frac{5}{8}$
B. $\frac{5}{12}$
C. $\frac{4}{5}$
D. $\frac{1}{6}$
E. $\frac{16}{21}$
F. 0
G. $\frac{11}{12}$
H. $\frac{2}{9}$
I. $\frac{9}{20}$
J. $\frac{11}{16}$
K. $\frac{11}{14}$
L. $\frac{7}{12}$

Page 20

A. $8\frac{3}{5}$, 7, $5\frac{3}{10}$, $6\frac{3}{4}$
B. $3\frac{1}{3}$, $5\frac{1}{3}$, $5\frac{4}{7}$, 1
C. $12\frac{9}{10}$, $9\frac{5}{12}$, 10, $12\frac{13}{30}$
D. $8\frac{5}{8}$, $3\frac{1}{4}$, $\frac{2}{3}$, $4\frac{3}{5}$

Page 21

A. $\frac{1}{5}$, $\frac{3}{28}$, $\frac{5}{36}$
B. $\frac{1}{42}$, $\frac{8}{27}$, $\frac{1}{7}$
C. $\frac{1}{2}$, $\frac{1}{9}$, $\frac{3}{20}$
D. $\frac{1}{5}$, $\frac{1}{6}$, $\frac{9}{40}$
E. $\frac{5}{72}$, $\frac{2}{45}$, $\frac{7}{12}$
F. $\frac{1}{4}$, $\frac{7}{24}$, $\frac{9}{40}$

Page 22

A. $\frac{7}{16}$
B. $\frac{2}{5}$
C. $\frac{5}{24}$
D. $\frac{3}{8}$
E. $\frac{4}{9}$
F. $\frac{7}{16}$
G. $\frac{1}{5}$
H. $\frac{2}{7}$
I. $\frac{4}{15}$
J. $\frac{1}{3}$
K. $\frac{3}{4}$
L. $\frac{1}{16}$
M. $\frac{1}{6}$
N. $\frac{1}{30}$

Page 23

A. $\frac{9}{10}$
B. $1\frac{2}{5}$
C. 5
D. $\frac{23}{63}$
E. $\frac{13}{20}$
F. $\frac{5}{9}$
G. $1\frac{1}{2}$
H. $1\frac{5}{6}$
I. $\frac{7}{12}$
J. $4\frac{2}{7}$
K. $2\frac{5}{8}$
L. $1\frac{1}{9}$
M. $1\frac{1}{3}$
N. $3\frac{1}{8}$

Page 24
A. $\frac{2}{9}$
B. $1\frac{3}{7}$
C. 8
D. 20
E. $\frac{4}{5}$
F. $2\frac{1}{2}$
G. 27
H. $4\frac{1}{2}$
I. $1\frac{1}{3}$
J. $6\frac{2}{3}$
K. $1\frac{1}{2}$
L. 12
M. $6\frac{2}{3}$
N. $2\frac{3}{4}$

Page 25
1. 0.07
2. 0.51
3. 0.176
4. 0.025
5. 0.9
6. 0.006
7. 0.32
8. 0.3
9. 0.8
10. 0.002
11. 0.01
12. 0.45
13. 0.604

Answer to riddle: JUMP OFF THE BOTTOM RUNG

Page 26
A. 0.9; 3.8; 15.4
B. 2.4; 0.2; 0.8
C. 1.1; 14.4; 7.5
D. 0.1; 0.3; 0.7
E. 0.37; 8.25; 0.62
F. 5.07; 9.13; 4.36
G. 0.88; 3.43; 12.09
H. 7.39; 15.66; 41.50
I. 4; 7; 3
J. 5; 10; 8
K. 12; 76; 47
L. 18; 30; 68

Page 27
A. 5.9; 4.33; 11.58; 18.15; 24.06
B. 5.5; 1.2; 39.05; 16.04; 16.71
C. 78.845; 23.882; 34.279; 13.817; 60.699
D. 7.1 (or 7.100); 9.226; 2.516; 6.808; 58.065
E. 44.87; 30.442; 63.083; 12.259; 35.232

Page 28
A. 0.7; 0.3; 0.8; 0.2
B. 0.49; 0.50 (or 0.5); 0.09; 0.01
C. 0.623; 0.705; 0.003; 0.014
D. $\frac{3}{10}$, $\frac{2}{5}$, $\frac{1}{2}$
E. $\frac{79}{100}$, $\frac{1}{2}$, $\frac{1}{4}$
F. $\frac{281}{1,000}$, $\frac{1}{4}$, $\frac{3}{200}$
G. >, >, =
H. >, <, <
I. =, <, >

Page 29
A. 4.2; 0.81; 8.4; 0.84; 9.9
B. 1.304; 0.045; 3.248; 60.8; 1.535
C. 2.24; 22.05; 1.44; 78.2; 103.32
D. 7.44; 0.744; 0.255; 0.286; 15.64

Page 30
A. 2.3; 10.5; 1.8; 6.8; 3.7
B. 0.8; 5.8; 1.23; 0.35; 2.16
C. 0.162; 1.96; 0.241; 2.881; 0.139
D. 0.63; 1.3; 2.4; 7.55; 0.051

Page 31
A. 4; 1.6; 5.1; 6.5; 4
B. 2.3; 3.9; 0.03; 0.06; 4.2
C. 2.3; 7; 8.1; 6; 0.32

Page 32
A. 8%
B. 60%
C. 17%
D. 51%
E. 23%
F. 85%
G. 33%
H. 92%
I. 18%; 13%
J. 36%
K. 67%

Page 33
A. $\frac{3}{100}$, $\frac{7}{100}$, $\frac{11}{100}$, $\frac{1}{2}$
B. $\frac{1}{10}$, $\frac{1}{5}$, $\frac{3}{20}$, $\frac{3}{50}$
C. $\frac{3}{4}$, $\frac{1}{4}$, $\frac{9}{20}$, $\frac{7}{10}$
D. $\frac{27}{50}$, $\frac{3}{10}$, $\frac{4}{5}$, $\frac{7}{20}$
E. 10%; 1%; 13%; 49%
F. 50%; 50%; 90%; 99%
G. 2%; 16%; 40%; 20%
H. 35%; 60%; 4%; 84%

Page 34
A. 0.09
B. 0.15
C. 0.48
D. 0.01
E. 0.37
F. 0.52
G. 0.66
H. 0.25
I. 0.74
J. 0.89
K. 0.04
L. 0.93
M. 38%
N. 99%
O. 72%
P. 44%
Q. 16%
R. 83%
S. 3%
T. 7%
U. 50%
V. 80%
W. 2%
X. 20%

Answer to riddle: BOTH HAVE PAGES

Page 35
A. 15
B. 13
C. 16
D. 9
E. 49
F. 30
G. 60
H. 120
I. 180
J. 51
K. 336
L. 390

Page 36
A. 70°
B. 35°
C. 108°
D. 130°
E. 45°
F. 110°
G. 120°
H. 95°

1. No. The sum of the angles of a triangle can't be greater than 180°.
2. Yes. The sum of the angles of a quadrilateral can be greater than 180°.

Page 37
A. P = 36 ft.
   A = 62 sq. ft.

B. P = 24 m
   A = 21 sq. m

C. P = 40 ft.
   A = 51 sq. ft.

D. P = 44 m
   A = 78 sq. m

E. P = 27 m
   A = 41.5 sq. m

F. P = 32 ft.
   A = 44 sq. ft.

Page 38
A. 20
B. 32.5
C. 80
D. 135
E. 120
F. 220
G. 60
H. 160
I. 180

Farmer Sprouts will buy F.

Page 39
A. 220
B. 96
C. 214
D. 232
E. 340
F. 249

The block with the greatest surface area is E.

Page 40
A. 72
B. 64
C. 125
D. 54
E. 48
F. 140
G. 100
H. 30
I. 42

Page 41
1. 3,000,605
2. 700,243
3. 2,162,999
4. 75,539,199
5. 618.5
6. 735.74
7. 32
8. 4,000
9. 80,900
10. 360,005

Page 42
1. 104,402
2. 2,832,991
3. 135,141
4. 245
5. 1.278
6. 4.2
7. $2\frac{7}{12}$
8. $\frac{7}{16}$
9. $1\frac{3}{8}$
10. $6\frac{2}{3}$

Page 43
1. 0.016
2. 0.04
3. 0.8
4. $\frac{1}{20}$
5. $\frac{1}{4}$
6. $\frac{2}{25}$
7. 39%
8. 90%
9. 85%
10. <
11. =
12. >
13. 10
14. 98
15. 260

Page 44
1. 108°
2. 58°
3. 40 sq. m
4. 15 sq. ft.
5. 32 ft.
6. 16 sq. m
7. 192 sq. in.
8. 120 cu. ft.